S0-EIL-782

Mining the Earth

John E. Young

Ed Ayres, Editor

Worldwatch Paper 109
July 1992

Library of Congress Catalog Number 92-061039
ISBN 1-878071-11-4

Printed on 100% recycled paper containing 15% post-consumer waste

Table of Contents

Introduction

Substances extracted from the earth—stone, iron, bronze—have been so critical to human development that historians name the ages of our past after them. But while scholars have carefully tracked human use of minerals, they have never accounted for the vast environmental damage incurred in mineral production.

Few people would guess that a copper mining operation has removed a piece of Utah seven times the weight of all the material dug for the Panama Canal. Few would dream that mines and smelters take up to a tenth of all the energy used each year, or that the waste left by mining measures in the billions of tons—dwarfing the world's total accumulation of more familiar kinds of waste, such as municipal garbage. Indeed, more material is now stripped from the earth by mining than by all the natural erosion of the earth's rivers.

Scouring the planet for its minerals has damaged large areas of land, often in remote, ecologically pristine areas. Mining projects now threaten 4 of every 10 national parks in tropical countries. The smelting of ores pumps millions of tons of sulfur dioxide and other pollutants into the atmosphere each year. Smelter pollution has created biological wastelands as large as 10,000 hectares, and accounts for a significant portion of the world's acid rain. The mineral industry's profligate use of energy makes it a substantial contributor to climate change as well as to more localized environmental problems.[1]

Yet in most discussions of threats to the global environment, mining is conspicuous only by its absence. The damage from mineral extraction is usually considered a local problem, and accepted—or imposed on local inhabitants—as an inevitable cost of economic development. It is also rarely tracked. For instance, the U.S. mining industry—though it is

I would like to thank Alyson Warhurst, Brian Skinner, and my colleagues at Worldwatch for comments on drafts of this paper.

clearly among the largest polluters—is not required to report its toxic emissions to state and federal regulators, as are most manufacturing industries. Cognizant that a country's overall prosperity usually correlates closely with its per capita use of mineral products, industrial nations have focused instead on the question of mineral supplies.

In the United States, for example, periodic waves of concern over future mineral supplies have led to the appointment of at least a half-dozen blue- ribbon panels on the subject since the twenties. Experts have assiduously questioned whether the country is going to have enough copper, tin, uranium, and other "nonrenewable resources." In 1978, a U.S. congressional committee requested a study whose title expressed the central question of virtually all these inquiries: Are we running out?[2]

Recent trends in price and availability suggest that for most minerals we are a long way from running out. Regular improvements in exploitive technology have allowed the production of growing amounts at declining prices, despite the exhaustion of many of the world's richest ores. For many minerals, much of the world has yet to be thoroughly explored.

In retrospect, however, the question of scarcity may never have been the most important one. Far more urgent is: Can the world afford the human and ecological price of satisfying its present appetite for minerals? If the answer is that it cannot, the challenge will be to find ways to continue developing and improving the quality of human life without constant growth in mineral extraction.

In turn, the question of what the world can afford depends on a true accounting of the costs of taking materials from the earth. Today's low mineral prices reflect only the immediate economics of extraction and distribution; they fail to consider the full costs of denuded forests, eroded land, dammed or polluted rivers, and the uprooting or decimation of indigenous peoples unlucky enough to live atop mineral deposits.

The environmental impacts of mineral extraction are particularly severe in developing countries, which produce a large portion of the world's mineral supplies but use a relatively small share. These nations also harbor some of the globe's greatest remaining concentrations of biological diversity. Mineral projects are among the largest causes of disturbance in such areas.

But while much of the damage is concentrated in the developing world, responsibility for most of it ultimately lies with those who use the most minerals—the fourth of humanity who live in industrial nations, enjoying material comforts others only dream of. The rich nations thus bear a special responsibility to help clean up the messes created to satisfy their needs, and to ensure that new damage is kept to a minimum.

In the long run, the most effective strategy for minimizing new damage is not merely to make mineral extraction cleaner, but to reduce the rich nations' needs for virgin (non-recycled) minerals. Hope for success lies in the economic maturity of today's most prosperous countries. Large quantities of minerals were required to build up their infrastructures—to make the concrete, steel, brick, and other materials needed for buildings and transportation systems. But once a society's basic structures are built, the quantities of additional materials it uses need not determine its quality of life. After a certain point, people's welfare may depend more on the caliber of a relatively small number of silicon microchips than on the quantities of copper, steel, or aluminum they use.

The sooner the whole world reaches such a point, the better. At the end of the minerals- and energy-intensive development path taken by today's industrial nations lies ecological ruin. Mining enough to supply a world expected to double in population during the next half century, with everyone using minerals at rates that now prevail in rich countries, would have staggering environmental consequences. Only by adopting a new development strategy—one that focuses on improving human welfare in ways that minimize the need for new supplies of minerals—can such consequences be averted.

Minerals in the Global Economy

Large-scale use of minerals began with the Industrial Revolution and grew rapidly for over two centuries. From 1750 to 1900, the world's overall use of minerals increased tenfold while population doubled.

Since 1900, it has jumped by at least thirteen-fold again.[3]

For some individual minerals, the growth has been even more extraordinary. Annual production of pig iron (the crude metal used to make steel) now stands at more than 500 million tons—or 22,000 times what it was in 1700. Outputs of copper and zinc are now 570 and 7,400 times greater than they were in 1800. Although aluminum was not commercially available until 1845, and was far too expensive for large-scale production until the modern electrolytic process was invented in 1886, smelters currently turn out 18 million tons of new metal each year.[4]

The term "minerals" encompasses a wide variety of substances taken from the earth. They are generally divided into four groups: metals, such as aluminum, copper, and iron; industrial minerals—such as lime and soda ash—that are valued for special qualities; construction materials, such as sand and gravel; and energy minerals, such as uranium, coal, oil, and natural gas (which, while they are extracted in large quantities at substantial cost to the environment, are outside the scope of this paper).[5]

Of the non-fuel minerals, stone, sand, and gravel are produced most widely and in the largest quantities. (See Table 1.) The principal use of the estimated 20 billion tons of such materials taken from the ground in 1991 was in construction, most often as ingredients of concrete. Ubiquitous in the earth's crust, they are generally used near the site where they are found.

Other nonmetals are more valuable and typically travel farther from mine to user. These include phosphates and potash (important ingredients in chemical fertilizers), lime (a major component of the cement that binds concrete), soda ash (an alkaline material used in many chemical processes), clays such as kaolin (an important ingredient in ceramics), and salt (most of which is used in the chemical industry, not in food).[6]

The most valuable minerals extracted are metals—of which the most important is iron. About thirty times as much iron is produced as the next most common metal. Steel—the cheap, strong material into which most iron is converted— costs one third to one half as much by weight as

Table 1. Estimated World Production of Selected Minerals, 1991

Mineral	Production[1] (thousand tons)
Metals	
Pig Iron	531,000
Aluminum	18,500
Copper	9,100
Manganese	6,700
Zinc	7,400
Chromium	3,800
Lead	3,370
Nickel	953
Tin	210
Molybdenum	110
Titanium	82
Tungsten	39
Cobalt	34
Cadmium	20
Silver	14
Mercury	6
Gold	2
Platinum-Group Metals	0.3
Nonmetals	
Stone	11,000,000
Sand and Gravel	9,000,000
Clays	500,000
Salt	186,000
Phosphate Rock	160,000
Potash	160,000
Lime	135,000
Gypsum	98,000
Soda Ash	33,000

[1]All data exclude recycling; figures for metals are smelter production or metal content of ore; figures for nonmetals are ore mined.

Sources: U.S. Bureau of Mines (USBM), *Mineral Commodity Summaries 1992* (Washington, D.C.: 1992); Donald G. Rogich, "Trends in Material Use: Implications for Sustainable Development," unpublished paper, Division of Mineral Commodities, USBM, April 1992; figures for stone, sand and gravel, and clays are Worldwatch estimates based on USBM, *Mineral Commodity Summaries.*

aluminum, its most common substitute. The estimated value of the steel sold each year is about five times the total for all other metals; and four of the nine metals immediately below iron on the production list—manganese, chromium, nickel, and molybdenum—are primarily used in steel-making.[7]

Aluminum is second to iron both in quantity and value of production. It is extremely important in aircraft construction because of its light weight, though in the United States, which leads all nations in aluminum consumption, the largest share of the metal is used to make beverage cans. Copper, the third-ranked metal in total tonnage, is primarily used as an electrical conductor. Zinc provides corrosion-resistant coatings for other metals; lead is used in electrical batteries and as an octane-boosting gasoline additive; and tin serves as a coating for steel cans.

The use of minerals is heavily concentrated in rich nations, and the disparities in use are most dramatic for metals. In 1990, the top eight industrial-nation users of aluminum, copper, and lead accounted for two thirds of world consumption. Eight or fewer wealthy countries took over half the iron ore and three fifths of zinc, tin, and steel supplies. A few decades earlier, these disparities were even greater; throughout the sixties, the industrial nations absorbed more than 80 percent of world steel production and at least 90 percent of other metals.[8]

Steady increases in demand fueled rapid expansion in minerals output until the seventies, when growth rates dropped substantially. From 1950 to 1974, use of eight economically important minerals grew between 2 and 9 percent each year, on average. By 1974-87, average growth rates for the same eight all fell under 2 percent. In the case of tin, use actually shrank.[9]

Five factors appear to underlie the slackening demand for minerals, and for metals in particular. Most involve basic changes in the economies of the chief users. First, industrial economies have grown more slowly since the oil crisis of 1973, so new construction has waned. Second, these economies are shifting away from heavy industry and toward services and high technology, so the amounts of physical materials needed are much smaller even when the economies do boom. The pharmaceutical and electronics industries, for example, are among the fastest-growing sectors in industrial economies, and are far less materials- and energy-intensive than traditional extractive and manufacturing industries.[10]

Third, recycling has reduced demand for metals (though not for other minerals, which are not easily recycled). In the United States, for instance, recycling provides a substantial share of consumption for many metals. (See Table 2.) Lead is recycled at a very high rate, largely because the metal is so toxic that its use is now tightly regulated. Seventy-three percent of 1990 U.S. lead consumption was supplied through recycling. Fifty-six percent of iron and steel consumption in the

Table 2. U.S. Metal Consumption and Recycling, 1990

Metal	Consumption (thousand tons)	Share of Consumption Provided by Recycling (percent)
Lead	1,297	73
Copper	2,168	60
Iron & Steel[1]	99,900	56
Gold	0.2	47
Aluminum	5,263	45
Platinum Group	0.1	45
Zinc[1]	1,060	43
Tin	45	38
Tungsten	8	29
Nickel	148	23
Cobalt[1]	7	22
Chromium	423	21
Silver[1]	5	14
Molybdenum	21	5

[1] 1989 figures

Source: Donald G. Rogich, "Trends in Material Use: Implications For Sustainable Development," unpublished paper, U.S. Bureau of Mines, Division of Mineral Commodities, April 1992.

United States now derives from scrap rather than fresh ore. Aluminum recycling is particularly widespread because its manufacture from scrap takes about one twentieth as much energy as its production from ore. Worldwide, nearly a third of the aluminum used each year is recycled.[11]

The fourth reason the demand for minerals, and especially for metals, has slackened is that new materials such as plastics, ceramics, and high-technology composites are now competing with metals—and are increasingly substituted for them—in many applications, from airplanes to construction. Glass fiber, for example, is supplanting copper in communications uses, and substitution of polyvinyl chloride pipes for copper ones effectively reduced yearly U.S. copper consumption by 13 percent in 1988.[12]

Fifth, and perhaps most important, the industrial nations have largely completed the building of their basic infrastructure of roads and buildings, and now need mineral products primarily to maintain or replace worn-out equipment and structures, not for massive new construction.

For all these reasons, minerals use is now growing faster in developing countries than in wealthier nations. Between 1977 and 1987, for example, the Third World's share of aluminum and copper use grew from 10 to 18 percent, and that of zinc from 16 to 24 percent. The increases are heavily concentrated in Mexico, Brazil, India, and the newly industrializing countries (NICs) of East Asia. Many products of traditional heavy industries, such as automobiles, are increasingly imported into industrialized nations from NICs—such as South Korea and Malaysia—so the mineral demand of those industries is partially shifting outside the major consumer nations.[13]

Production of minerals is more widely distributed through the world than consumption, although for some individual minerals there are exceptions. For instance, known deposits of cobalt, chromium, and the platinum-group metals are concentrated in a few countries, making supplies vulnerable to local political developments.[14]

Production is declining, however, in many of the countries that are major minerals users. West European countries, for instance, have depleted their high-grade mineral reserves and now rely mostly on imports. Japan imports virtually all the minerals it uses. The United States is still an important producer of many minerals, but gets nearly all of its bauxite and alumina (the refined bauxite used by aluminum smelters), three fourths or more of its nickel, chromium, and tin, and about a third of its zinc from foreign sources.[15]

Despite a long history of mining, the republics of the former Soviet Union still produce large amounts of minerals. Australia and South Africa are major mineral sources, and the list of important producers also includes such developing nations as Brazil, Chile, China, and Zaire. (See Table 3.)

Each major industrial region looks to a corresponding part of the Third World for most of its mineral imports: the United States to Latin America, Western Europe to Africa, and Japan to Asia and Oceania. Mineral trade relationships also often reflect old colonial ties. For instance, Zaire supplies Belgium, its former ruler, with about two thirds of its imported copper, and Zambia supplies Britain with a third of its imports of the same metal.[16]

Mineral "reserves"—an often misunderstood concept—have been a subject of abiding interest for the world's great mineral-using nations. Reserves consist of deposits whose existence has been documented by detailed surveying and that are judged to be minable at a cost no higher than current market prices. At current use rates, global reserves of economically important minerals range from 20-30 years of supply (lead, tin, and zinc) to 200 years (bauxite). Although 20 years may seem alarmingly short, there is little danger of the world soon running out. Mineral resources—deposits whose presence is indicated by preliminary surveys or other geologic evidence but that are not yet economically viable—are far greater than reserves, and exploration constantly moves deposits from the resources to the reserves category. In recent decades, mineral reserves have generally grown at least as fast as production.[17]

Many of the best reserves now lie in developing countries, since industrial nations have a much longer history of mining. The Commonwealth of Independent States and the other former members of the Warsaw Pact

Table 3. Major Mineral-Producing Countries, 1991

Mineral	Countries	Share in World Production
		(percent)
Bauxite	Australia	38
	Guinea	16
Chromium	South Africa	35
	Soviet Union/CIS[1]	30
Cobalt	Zaire	50
	Zambia	21
Copper	Chile	20
	United States	18
Gold	South Africa	29
	United States	15
Iron Ore	Soviet Union/CIS	24
	Brazil	17
Lead	Australia	16
	United States	14
Manganese	Soviet Union/CIS	38
	China	14
Molybdenum	United States	54
	Chile	13

Mineral	Countries	Share in World Production (percent)
Nickel	Soviet Union/CIS	27
	Canada	21
Phosphate Rock	United States	29
	Soviet Union/CIS	23
Platinum Group	South Africa	48
	Soviet Union/CIS	44
Potash	Soviet Union/CIS	32
	Canada	28
Silver	Mexico	15
	Peru	13
Tin	China	19
	Brazil	15
Tungsten	China	52
	Soviet Union/CIS	21
Zinc	Canada	16
	Australia	14

[1] Commonwealth of Independent States (successor to the Soviet Union in 1991).

Source: U.S. Bureau of Mines, *Mineral Commodity Summaries 1992* (Washington, D.C.: 1992).

also possess large reserves of many important minerals, including iron ore, manganese, chromium, and nickel, and their potential resources are enormous. However, economic and political turmoil in the former socialist countries makes them unlikely sites for new mineral projects, which usually require large capital investments and long lead times before production can begin.[18]

Overall, scarcity of mineral deposits does not appear likely to constrain the production of most important minerals in the foreseeable future. Much more probable, however, are reductions in output due to environmental concerns.

Laying Waste

Mining is the original dirty industry. As the German scholar Georgius Agricola put it in his 1550 treatise on mining: "The fields are devastated by mining operations...the woods and groves are cut down, for there is need of an endless amount of wood for timbers, machines, and the smelting of metals. And when the woods and groves are felled, then are exterminated the beasts and birds.... Further, when the ores are washed, the water which has been used poisons the brooks and streams, and either destroys the fish or drives them away."[19]

Four centuries later, mining's environmental effects remain much the same, but on a vastly greater scale. Modern machinery can do in hours what it took men and draft animals years to do in Agricola's time. Larger equipment reflects the growing scale of the industry. A typical truck used in hard-rock mining in 1960 weighed 20-40 tons, for example; in 1970 it weighed in at 80-200 tons. The size of the shovels used to move ore increased from 2 to 18 cubic meters over the same period. Such technological advances allowed world mineral production to grow rapidly—and proportionately increased the harm to the environment.[20]

Mining and smelting have created large environmental disaster areas in many nations. (See Table 4.) In the United States, which has a long history of mining, at least 48 of the 1,189 sites on the Superfund hazardous-waste cleanup list are former mineral operations. The largest Superfund site stretches across the state of Montana, along a 220-kilometer stretch of

Table 4. Environmental Impacts of Selected Mineral Projects

Location/Mineral	Observation
Ilo-Locumbo area, Peru copper mining and smelting	The Ilo smelter emits 600,000 tons of sulfur compounds each year; nearly 40 million cubic meters per year of tailings containing copper, zinc, lead, aluminum, and traces of cyanides are dumped into the sea each year, affecting marine life in a 20,000-hectare area; nearly 800,000 tons of slag are also dumped each year.
Nauru, South Pacific phosphate mining	When mining is completed—in 5-15 years—four fifths of the 2,100-hectare South Pacific island will be uninhabitable.
Pará state, Brazil Carajás iron ore project	The project's wood requirements (for smelting of iron ore) will require the cutting of enough native wood to deforest 50,000 hectares of tropical forest each year during the mine's expected 250-year life.
Russia, Severonikel smelters	Two nickel smelters in the extreme northwest corner of the republic, near the Norwegian and Finnish borders, pump 300,000 tons of sulfur dioxide into the atmosphere each year, along with lesser amounts of heavy metals. Over 200,000 hectares of local forests are dying, and the emissions appear to be affecting the health of local residents.
Sabah Province, Malaysia Mamut Copper Mine	Local rivers are contaminated with high levels of chromium, copper, iron, lead, manganese, and nickel. Samples of local fish have been found unfit for human consumption, and rice grown in the area is contaminated.
Amazon Basin, Brazil gold mining	Hundreds of thousands of miners have flooded the area in search of gold, clogging rivers with sediment and releasing an estimated 100 tons of mercury into the ecosystem each year. Fish in some rivers contain high levels of mercury.

Source: Worldwatch Institute, based on sources documented in note 21.

Silver Bow Creek and the Clark Fork River. Water and sediments in the river and a downstream reservoir are contaminated with arsenic, lead, zinc, cadmium, and other metals, which have also spread to nearby drinking-water aquifers. Soils throughout the local valley are contaminated with smelter emissions.[21]

The Clark Fork Basin was the site of more than 100 years of mining and smelting, including what was at one time the largest open pit in the world, the Berkeley Pit copper mine. The pit and a network of underground mine workings contain more than 40 billion liters of acid mine water that rises a little higher each year, threatening local aquifers and already-tainted streams with contamination. The Clark Fork Coalition, a local environmental group, estimates that cleaning up the pit and other sites in the area could cost over $1 billion. A proposed large new copper mine in the Cabinet Mountains area of northwest Montana now endangers another section of the Clark Fork's drainage.[22]

The environmental damage done in producing a particular mineral is determined by such factors as the ecological character of the mining site, the quantity of material moved, the depth of the deposit, the chemical composition of the ore and the surrounding rocks and soils, and the nature of the processes used to extract purified minerals from ore. (See Table 5.) Damage varies dramatically with the type of mineral being mined. For example, stone ranks first in production, but its extraction probably causes less overall harm than that of several metals. Since stone and other construction materials are usually taken from shallow or naturally exposed deposits and used with little or no processing, the environmental impacts are mostly limited to land disturbance at the quarry or gravel pit, and relatively few wastes are generated.

At the other end of the damage spectrum, metals are produced through a long chain of processes, each of which involves pollution and the generation of waste. Copper production, for instance, typically involves five stages. First, soil and rock (called overburden) that lie above the ore must be removed. The ore is then mined, after which it is crushed and run through a concentrator, which physically removes impurities. The concentrated ore is reduced to crude metal at high temperatures in a smelter, and the metal is later purified, through remelting, in a refinery.

Table 5. Environmental Impacts of Minerals Extraction

Activity	Potential Impacts
Excavation and Ore Removal	• Destruction of plant and animal habitat, human settlements, and other surface features (surface mining) • Land subsidence (underground mining) • Increased erosion; silting of lakes and streams • Waste generation (overburden) • Acid drainage (if ore or overburden contain sulfur compounds) and metal contamination of lakes, streams and groundwater
Ore Concentration	• Waste generation (tailings) • Organic chemical contamination (tailings often contain residues of chemicals used in concentrators) • Acid drainage (if ore contains sulfur compounds) and metal contamination of lakes, streams, and groundwater
Smelting/Refining	• Air pollution (substances emitted can include sulfur dioxide, arsenic, lead, cadmium and other toxic substances) • Waste generation (slag) • Impacts of producing energy (most of the energy used in extracting minerals goes into smelting and refining)

Source: Worldwatch, compiled from various sources.

Most of today's mines are surface excavations rather than underground complexes of tunnels and shafts, so the miner's first task is to remove whatever lies over a mineral deposit—be it a mountain, a forest, a farmer's field, or a town. For any given mineral, surface mining produces more waste than working underground. In 1989, U.S. surface mines produced 8 times as much waste per ton of ore as underground mines did. That same year, overburden accounted for more than a third of the 3.4 billion tons of material handled at non-fuel mines. Such material, while it may be chemically inert, can clog streams and cloud the air over large areas. If the overburden contains sulfur compounds—common in rock containing metal ores—it can react with rainwater to form sulfuric acid, which then may contaminate local soils and watercourses.[23]

Similar but more severe effects often stem from extraction of the ore itself and from the disposal of tailings, the residue from ore concentration. Up to 90 percent of metal ore ends up as tailings, which are commonly dumped in large piles or ponds near the mine. The finely ground material makes contaminants that were formerly bound up in solid rock (such as arsenic, cadmium, copper, lead, and zinc) accessible to water. Acid drainage, which exacerbates metal contamination, is often a problem, since sulfur makes up more than a third of the commonly mined ores of many metals, including copper, gold, lead, mercury, nickel, and zinc. Tailings also usually contain residues of organic chemicals—such as toluene, a solvent damaging to human skin and to the respiratory, circulatory, and nervous systems—that are used in ore concentrators. Ponds full of tailings cover at least 3,500 hectares in the Clark Fork area and 2,100 hectares at the Bingham Canyon copper mine in Utah.[24]

A particularly dramatic example of the impact of tailings disposal is the Panguna copper mine on Bougainville, an island in Papua New Guinea that since mid-1989 has been controlled by secessionist rebels. Before it was closed, the mining operation dumped 600 million tons of metal-contaminated tailings—130,000 tons each day—into the Kawerong River. The wastes cover 1,800 hectares in the Kawerong/Jaba river system, including a 700-hectare delta at its mouth, 30 kilometers from the mine. Environmental writer Don Hinrichsen described the Jaba River as "so full of sediments from the Bougainville Copper Mine that its slate-grey waters are completely dead....Wading into the river to take samples is like inching through moving mud." Local anger at the destruction of the area by mining was a major cause of the civil war.[25]

Smelting, the next stage of the extraction process, can produce enormous

"Up to 90 percent of metal ore ends up as tailings, which are commonly dumped in large piles or ponds."

quantities of air pollutants. Worldwide, smelting of copper and other nonferrous (non-iron) metals releases an estimated 6 million tons of sulfur dioxide into the atmosphere each year—8 percent of total emissions of the sulfur compound that is a primary cause of acid rain. Nonferrous smelters can also pump out large quantities of arsenic, lead, cadmium, and other heavy metals. If they lack pollution control equipment, aluminum smelters emit tons of fluoride, which can concentrate in vegetation and kill not only the plants but, in some cases, animals that eat them.[26]

Uncontrolled smelters have produced some of the world's best-known environmental disaster areas—"dead zones" where little or no vegetation survives. Such an area around the Sudbury, Ontario, nickel smelter in Canada measures 10,400 hectares; acid fallout from the smelter has destroyed fish populations in lakes 65 kilometers away. Between 1896 and 1936, a smelter at Trail, British Columbia killed virtually all conifers within 19 kilometers and retarded tree growth up to 63 kilometers away. In the United States, a dead zone surrounding the Copper Hill smelter in Tennessee covers 7,000 hectares. In the United Kingdom, 400,000 hectares of agricultural land have been lost to metal smelting since Roman times; and in Japan, about 6,700 hectares of cropland are too contaminated for rice production.[27]

New dead zones, such as the area surrounding the Severonikel nickel smelter in Russia, are still being created. Smelters in industrial countries are now often required by law to have pollution control equipment, but few in developing countries or the formerly socialist nations have any such controls. For each kilogram of copper produced, 12.5 times more sulfur dioxide is released to the air from Chilean smelters than from those in the United States.[28]

The grade of an ore—its metal content in percentage terms—is a critical factor in determining the overall impact of metal mining. The average grade of copper ores, for example, is lower than that for any of the other major metals. Four centuries ago, copper ores typically contained about 8 percent metal; the average grade of ore mined now is under 1 percent. One consequence of the drop in grade is that more than eight times as much ore now must be processed to obtain the same amount of copper.

An estimated 990 million tons of ore were mined to produce about 9 million tons of copper in 1991.[29] (See Table 6.)

Even this figure understates the total amount of material moved, since it does not include overburden. The scale of the industry is apparent, however, in the size of the holes it creates. Some 3.3 billion tons of material—seven times the amount moved for the Panama Canal—have been taken from Utah's Bingham Canyon copper mine. Now 774 meters deep, this mine is the largest human excavation in the world. It gained

Table 6. Estimated Ore Production, Average Grade, and Waste Generation, Major Minerals, 1991

Mineral	Ore	Average Grade	Waste
	(million tons)	(percent)	(million tons)
Copper	1,000	0.91	990
Gold	620	0.00033	620
Iron	906	40.0	540
Phosphate	160	9.3	140
Potash	160	17.0	130
Lead	135	2.5	130
Aluminum/Bauxite	109	23.0	84
Nickel	38	2.5	37
Tin	21	1.0	21
Manganese	22	30.0	16
Tungsten	15	0.25	15
Chromium/Chromite	13	30.0	9
Total	3,200		2,700

Waste figures do not include overburden.
Totals do not add due to rounding.

Sources: Worldwatch Institute, based on production estimates in U.S. Bureau of Mines, *Mineral Commodity Summaries 1992* (Washington, D.C.: 1992), and grade estimates in Donald G. Rogich, "Trends in Material Use: Implications for Sustainable Development," unpublished paper, Division of Mineral Commodities, U.S. Bureau of Mines, April 1992.

another distinction in 1987 when its operator, Kennecott Copper, inadvertently reported its toxic chemical releases to the Environmental Protection Agency's Toxics Release Inventory (a national toxic-chemicals reporting system from which the mining industry is exempt). Out of the 18,000 industrial facilities reporting, the Bingham Canyon mine ranked fourth in total toxic releases and first in metals. The company discontinued reporting the following year, but the scale of its releases spurred legislative efforts in 1991 and 1992 to include the mining industry in the inventory in the future.[30]

Gold mining also requires the processing of large amounts of material, since the metal occurs in concentrations best measured in parts per million. An estimated 620 million tons of waste are produced in gold mining each year— even more than is produced in iron mining, which yields 26,000 times as much metal by weight. The operators of the Goldstrike mine in Nevada—the largest in the United States—each day move 325,000 tons of ore and waste to produce under 50 kilograms of gold. In Brazil's Amazon Basin, thousands of small-scale gold miners are using a technique called hydraulic mining to extract as much as 120 tons of gold per year. This involves blasting gold-bearing hillsides with high-pressure streams of water, and then guiding the water and sediment through sluices that separate tiny amounts of gold, which is heavier, from tons of non-valuable material, which then pollutes local rivers. The technique is so environmentally destructive that its use was halted over 100 years ago in California, where it did widespread damage during the state's legendary gold rush.[31]

Since 1979, when the price of gold soared to an all-time high of $850 per ounce, a gold rush has swept the world. Waves of gold seekers have invaded remote areas in Brazil, other Amazonian countries, Indonesia, the Philippines, and Zimbabwe. Dramatic environmental damage has resulted. Hydraulic mining has silted rivers and lakes, and the use of mercury—an extremely toxic metal that accumulates in the food chain and causes neurological problems and birth defects—to capture gold from sediment has contaminated wide areas. Miners release an estimated 100 tons of mercury into the Amazon ecosystem each year. An estimated 32 tons are released each year in the watershed of the Madeira River (a major Amazon tributary) alone. Mercury levels in most carnivorous and some omnivorous fish in the Madeira exceed the maximum safe levels for human consumption set by many nations.[32]

In North America, heap leaching, a new technology that allows gold extraction from very low-grade ores, is now in wide use. Miners spray cyanide solution, which dissolves gold, on piles of crushed ore or old tailings. After repeated circulation through the ore, the liquid is collected and gold is extracted from it. Both cyanide-solution collection reservoirs and the contaminated tailings left behind after leaching pose hazards to wildlife and groundwater. In October 1990, for instance, 38 million liters of cyanide solution spilled from a reservoir at the Brewer Gold Mine, near Jefferson, South Carolina, into a tributary of the Lynches River. The spill, caused by a dam break after a heavy rain, killed as many as 10,000 fish. Thousands of birds also die each year when they mistake cyanide impoundments for lakes.[33]

Fossil-fuel-powered machinery has allowed mining to expand to such a degree that its effects now rival the natural processes of erosion. An estimated 24 billion tons of non-fuel minerals are taken from the earth each year, of which about 2.7 billion tons are waste (not including overburden). Taking overburden into account, the total amount of material moved is probably at least 28 billion tons—about 1.7 times the estimated amount of sediment carried each year by the world's rivers.[34]

An estimated half-million hectares of land—including mines, waste disposal sites, and areas of subsidence over underground mines—are directly disturbed by non-fuel mining each year. Most of this land will bear the scars indefinitely. Historian Elizabeth Dore, describing the effects of 500 years of mining on the Bolivian landscape, writes: "Silver and tin are gone; in their place rise mountains of rock, slag, and tailings.... Saturated with mercury, arsenic, and sulfuric acid, the iridescence of these rubbish heaps provides a psychedelic reminder of the past." The damage is not limited to the mine site. As Dore puts it, mining initiates "a chain of soil, water, and air contamination" that can alter the ecosystems of large areas.[35]

Moving billions of tons of material and crushing and melting rock requires large amounts of energy, and supplying it can cause major dam-

"An estimated 24 billion tons of non-fuel minerals are taken from the earth each year, of which about 2.7 billion are waste."

age to local ecosystems. Ever since Agricola's time, for example, wood-fired smelters have threatened nearby forests. In southern England, the Sussex iron industry was effectively wiped out when it destroyed the local woods that provided its fuel supply. In the late nineteenth century, more than 2 million cords of wood were used as smelter fuel in Nevada's Comstock Lode—described by one observer as "the tomb of the forests of the Sierras."[36]

Today, demand for energy to extract and process minerals is playing a major role in the deforestation and inundation of large parts of the Amazon Basin. A huge iron ore mining and smelting project at Carajás, in the Brazilian state of Pará, threatens a large area of tropical forest. The project's 20 planned pig-iron smelters will need an estimated 2.4 million tons of charcoal each year, which if produced from native trees will require an estimated 50,000 hectares of forest to be logged annually. According to ecologist Philip Fearnside, high costs make it unlikely that plantations will supply much of the wood, and the state enterprise that owns the project has thus far done little to develop plantation production. The mine is expected to operate for 250 years.[37]

The iron ore facilities are only one piece of Brazil's colossal Grande Carajás Project, a vast state-run development scheme that also includes bauxite, copper, chromium, nickel, tungsten, tin, and gold mines; mineral processing plants; hydroelectric dams; deep-water ports; and other enterprises. Aluminum smelters, including the 330,000-ton-per-year Albrás plant (another element of the Carajás project) now take most of the electricity output of the enormous—and enormously destructive—Tucuruí hydroelectric station. Albrás, a major justification for the dam's construction, receives power at one fourth the cost of generation (and one third the average cost of Brazilian electricity). Aluminum smelting took 12 percent of Brazilian electricity in 1988, and the industry's power requirements more than doubled between 1982 and 1988.[38]

Aluminum production is particularly energy-intensive. Unlike most other metals, which can be obtained by simply heating the ore, aluminum forms such tight chemical bonds that it can only be economically extracted through a process involving the direct application of electrical current. Modern aluminum smelters require 13-18 kilowatt-hours of electricity to produce a kilogram of metal. The world aluminum industry uses an estimated 290 billion kilowatt-hours of electricity each year—more than is used for all purposes on the entire African continent. Additional energy is used in mining bauxite, and in processing it into the alumina that is smelted. All told, aluminum production requires an estimated 3.8 billion gigajoules (GJ) of energy each year—around 1 percent of world energy use. Much of it is purchased at unusually low rates—subsidized by governments at heavy human and environmental expense.[39]

Though figures are sparse, the mineral industry as a whole is clearly among the world's largest users of energy, and thus a major contributor to the impacts of energy use, including climate change. While aluminum is the most energy-intensive of the metals, steel and copper are also large energy-users. Steelmaking, in fact—because of its sheer volume—is probably the largest energy user of all mineral industries; in the United States, which produces only 10 percent of the world's supply, steelmaking required 2.2 billion GJ in 1988. Worldwide, copper production takes about 1 GJ. All told, the minerals industry probably accounts for 5 to 10 percent of world energy use.[40]

The efficiency of energy use in smelting and refining metals has improved over time. Today's U.S. aluminum smelters, for example, use between half and two thirds as much electricity as those built in the late forties. Some new copper smelting technologies use only about 60 percent as much energy as traditional methods. But while smelting has improved, long-term trends in ore grades and accessibility of deposits tend to increase the energy used per unit of metal mined. Declining ore grades increase energy needs, because more ore must be mined, greater quantities of waste material must be handled, and more effort is required to concentrate and smelt the ore. And as more remote, deeper deposits are mined to replace those more easily reached, more energy is required in order to dig bigger holes and transport the ore longer distances.[41]

Another, often forgotten side of the mining industry is its effects on local people and their environment. The rush to produce more minerals and gain export revenue—especially in countries which have few other major sources of income—has had devastating consequences for those

"The rush to produce more minerals has had devastating consequences for those whose homelands lie over mineral deposits."

whose homelands lie over mineral deposits. Developers and funders of large mining projects have rarely considered the future of local people during project planning—or when deciding whether to proceed with a project in the first place. The Panguna copper mine in Papua New Guinea, where massive local environmental damage led to a civil war, is a particularly dramatic example. Local people received little compensation for the confiscation or destruction of their land. But the mine was a huge source of income for the national government; before the rebels succeeded in closing it in 1989, it was yielding 17 percent of the country's operating revenue and 40 percent of its export income.[42]

Indigenous people, in particular, are often simply pushed aside in the rush to mine valuable deposits. In the Amazon, for example, the Yanomami, an interior tribe who avoided contact with the outside world until recent decades, are only the most recent case of a people decimated for the rocks beneath their feet. Thousands of miners crowded their homeland in the northern Brazilian states of Roraima and Amazonas during the eighties, polluting local rivers with sediment and mercury and bringing unfamiliar diseases to the indigenous population. It is estimated that at least 15 percent of the Yanomami have died from malaria.[43]

The Brazilian government was for many years unwilling to do anything substantial to save the Yanomami, and local and state officials have long been active supporters of the area's mining operations. As the then-governor of Roraima put it in 1975, "an area as rich as this cannot afford the luxury of conserving half a dozen Indian tribes who are holding back the development of Brazil." Fortunately, in 1990 and 1991, under heavy pressure from human rights organizations and other governments, President Fernando Collor de Mello sent in federal police to expel miners from Yanomami lands, and late in 1991, he signed an executive decree establishing a 9 million hectare reserve (in which mining was to be banned) for the tribe.[44]

At What Cost?

While mineral prices have fluctuated dramatically, the overriding trend in recent decades has been downward, with most of the plunge occurring during the dozen years following the 1973 oil crisis. After adjusting for inflation, the International Monetary Fund (IMF) index of nonfuel mineral prices declined by half between 1974 and 1986. (See Figure 7-1.)

Real prices recovered somewhat in the late eighties, but have never returned to the levels of the fifties, sixties, and early seventies. They were again in decline in 1990, 1991, and early 1992.

Why are mineral prices so low? One reason is that many nations subsidize development of their domestic mineral resources. Since the twenties, for example, the United States has offered mining companies generous tax exemptions or "depletion allowances." Miners can deduct from 5 to 22 percent of their gross income, depending on the mineral. Unlike conventional depreciation, depletion allowances are not based on capital investments made by the company. In fact, the allowances may be taken for as long as the mine operates—even after the company's

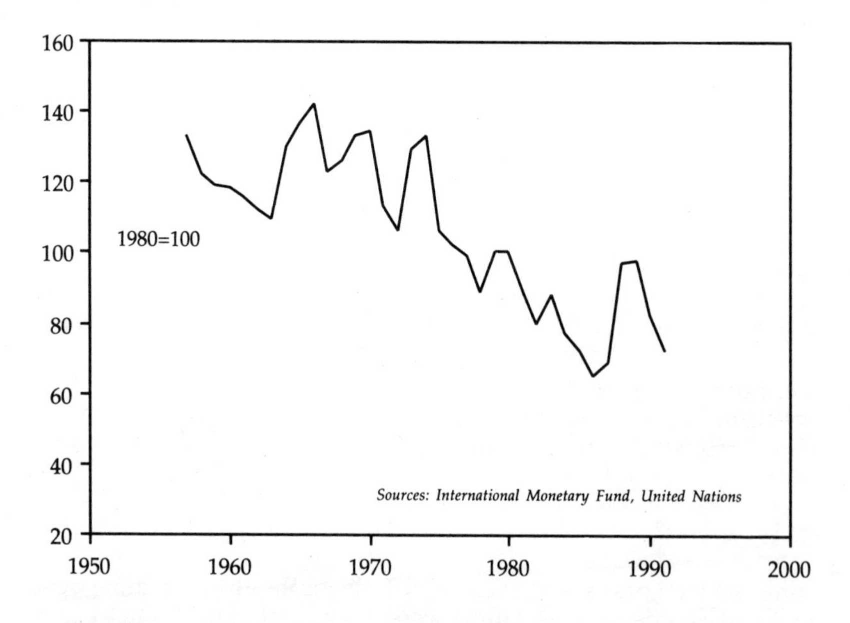

Figure 1. Index of real non-fuel mineral prices, 1957–91

investment is fully recovered. In addition, mining companies may also deduct much of the cost of exploring for and developing mineral deposits.[45]

The lost taxes added up to a $5 billion subsidy to the U.S. mining industry over the last decade. The President's budget projects that the 1992 subsidy will be $560 million. According to the Joint Committee on Taxation of the U.S. Congress, the Treasury could gain $2 billion over the next five years if the mineral industry were taxed on the same basis as others.[46]

The U.S. mining industry receives another large but uncalculated subsidy through virtual giveaways of federal land under the General Mining Act of 1872. This legal relic of the frontier era allows those who find hard-rock minerals (such as gold, silver, lead, iron, and copper) in public territory to buy the land for $12 per hectare or less. Since the government retains no rights to the land, the Treasury does not receive royalties on minerals taken from it, as it does from oil, gas, and coal found on public land. Even very low royalty levels would yield large sums, since miners took $4 billion worth of hard-rock minerals from former federal lands in 1988.[47]

Japan offers loans, subsidies, and tax incentives for exploration and development of domestic mineral deposits. Similarly, the French government offers financial assistance for minerals exploration, and also makes direct investments in mineral projects through the Bureau de Recherches Geologiques et Minières (BRGM), a state-owned enterprise. Germany is considerably less generous, but does offer direct support for exploration. Most industrial nations must now look abroad for new mineral supplies, however. Japan's domestic mineral resources are, by the geological luck of the draw, quite limited. Western Europe was once quite rich in minerals, but demand during two centuries of industrial development has depleted most major deposits.[48]

Industrial nations have thus also tried to ensure continued access to cheap minerals supplies through their international trade and aid policies. The Japanese government subsidizes foreign mineral projects of Japanese companies through low-interest loans, loan guarantees, and direct government investments. The BRGM helps French companies with funding for exploration and development of overseas projects. Germany offers investment guarantees, minimum rate-of-return guarantees, and favorable loans for foreign mineral investments by its companies. These nations have often joined the United States in supporting efforts by development institutions, including the World Bank, to finance mineral projects in developing countries—at times with the explicit intention of securing future minerals sources.[49]

Historically, nations have justified subsidies for minerals extraction on national security grounds. Minerals are critical to arms production, and supplies have often been equated with military power. Progress in the arms race between the United Kingdom and Germany prior to World War I was measured in steel production. The aluminum industry would not be what it is today were it not for the many forms of special support—especially cheap electric power—given by national governments that during World War I came to understand the metal's importance in modern warfare. The role of aluminum in World War II was summed up by a U.S. analyst in 1942 with the following equation: "Electric power ➔ aluminum ➔ bombers ➔ victory."[50]

Today, powerful political and economic interests support continued subsidies for domestic mineral production. While the mining industry and manufacturers who purchase mineral products benefit from this practice, taxpayers end up with the bill—and markets are skewed toward use of virgin rather than recycled materials. One analyst estimated in 1977 that U.S. tax subsidies for virgin materials—including wood products as well as minerals—made such items about 10 percent cheaper than they would be otherwise. The actual impact of subsidies on virgin materials prices is probably much greater, since other, unquantified subsidies were not taken into account.[51]

Given the growing dependence of rich nations on foreign mineral supplies and their willingness to assist with the development of these resources, developing nations would seem well positioned to benefit from their mineral wealth. But the people of most mineral-exporting developing countries have gained little from mining. Take the case of Bolivia, as described by writer David J. Fox: "The world knows Bolivia,

"Many developing nations seem to have been dragged down economically by their dependence on revenue from mineral exports."

if it knows it at all, as a mining country.... [It] is a member of a small group of African and Latin American countries for whom mining is the cornerstone of their economies; no other country of this group...has been so dependent on mining over such a long period as has Bolivia. But...great mining wealth has brought only poverty to the average Bolivian; in no other country in South America is the standard of living so low." Bolivia is not alone. Indeed, many developing nations seem to have been dragged down economically by their dependence on revenue from mineral exports.[52]

At least fourteen developing countries get a third or more of their export revenues from minerals. (See Table 7.) For these nations, the effect of nearly two decades of falling prices has been less foreign exchange with which to purchase the manufactured goods—from tractors to televisions to pharmaceuticals—they import. Reduced foreign exchange earnings have also made it difficult or impossible for many minerals producers to repay their international debts. They are among the world's most indebted nations. On average, the developing countries listed in the table bear external debts 1.4 times greater than their gross national product (GNP). By comparison, the average ratio of debt to GNP among the nations the World Bank classifies as severely indebted is 0.6.[53]

Zambia provides a dramatic example of what can happen when a nation gets trapped between heavy debt and falling prices for its main export. The country went into an economic tailspin when the price of copper, which provides 86 percent of its export revenue, dropped sharply in the early eighties. The results were calamitous: twice as many Zambian children died from malnutrition in 1984 as in 1980. A series of IMF-prescribed recovery plans—which cut domestic spending and devalued the national currency in an attempt to boost mineral exports and reduce the external debt—have failed to rescue Zambia from its economic slide. As of 1989, Zambia's debt was 1.4 times as great as its GNP.[54]

The current plight of Third World minerals exporters is the product of a collision between economic development strategies and unforeseen trends in the mining industry and minerals use. Unfortunately, the projected positive trends in minerals demand and prices that were the foun-

Table 7. Share of Minerals in Value of Total Exports, Selected Countries, Recent Years[1]

Country	Mineral(s)[2]	Share (percent)
Botswana	diamonds, copper, nickel	87
Zambia	copper	86
Zaire	copper, diamonds	71
Suriname	bauxite/alumina, aluminum	69
Papua New Guinea	copper	62
Liberia	iron ore, diamonds	60
Jamaica	bauxite/alumina	58
Togo	phosphates	50
Central African Republic	diamonds	46
Mauritania	iron ore	41
Chile	copper	41
Peru	copper, zinc, iron ore, lead, silver	39
Bolivia	zinc, tin, silver, antimony, tungsten	35
Dominican Republic	ferronickel	33
Guyana	bauxite	31
South Africa	gold	29

[1]Figures are for most recent year available; most are 1990 or 1991; earliest (Zaire) is for 1986.
[2]Minerals are listed in order of total export value.

Source: International Monetary Fund, *International Financial Statistics,* May 1992.

dation of the economic plans of many exporters during the sixties and seventies have vanished in the last decade.

Throughout the fifties, privately owned companies essentially controlled— through loose oligopolies—the prices of most major minerals, owned most major projects, and garnered the bulk of the profits. Although cyclical swings were not eliminated, prices were maintained at profitable levels by the combination of steadily increasing demand and industry control. Companies cut production during economic slumps in the major consumer nations, and anticipated economic upturns with investments in new capacity. The power of private firms made it very difficult for Third World governments to impose substantial local taxation or royalty requirements on foreign-owned mineral firms. If they did, the mining companies—which often owned not only mines but also smelting and refining facilities—could keep local tax bills to a minimum by selling their ores to their own processing subsidiaries at low prices, a practice known as transfer pricing.[55]

But in the tide of nationalism of the sixties and seventies, many governments—including newly independent nations and some long-independent ones just learning to flex their muscles—tried to increase their share of the profits from mining through higher taxes, new royalty requirements, and measures against transfer pricing. Some of them, including Bolivia, Chile, Guinea, Guyana, and Zaire, imposed minimum local ownership requirements or nationalized foreign-owned mines. Inspired by the example of the Organization of Petroleum-Exporting Countries, producers of some commodities, including bauxite, copper, and iron ore, banded together in attempts to increase prices.[56]

At the same time, the World Bank and other international development institutions promoted an economic strategy that emphasized the use of mining revenue as working capital for industrialization and overall economic growth. Optimistic mineral price projections fueled a rush of investment in mining projects. An increasing number were undertaken by state-owned enterprises rather than private firms, so national governments ended up bearing the debt. Although World Bank investments in mineral projects were not exceptionally large compared with other economic sectors, its contributions leveraged considerable funding from other sources. For example, the Bank contributed only $60 million of the $501 million required for the Guelbs iron ore project in Mauritania, with the balance coming from European, Arab, and Japanese institutions. Private banks, awash in billions of dollars amassed by the newly wealthy oil-exporting countries, also plowed newfound capital into mineral projects.[57]

Minerals projects are expensive. They are also often connected more closely to the outside world than to the host country. For instance, the

Guelbs project—which consists of an iron mine and ore concentrator, a port, and a railroad to connect them—has virtually no links with the rest of Mauritania's economy, which it dwarfs. When financing was arranged in 1979, the investment required for the first phase of the project topped the nation's gross domestic product, which then stood at $470 million.[58]

As long as future demand appeared strong, however, and prices remained fairly high, such projects seemed viable. Mineral export revenues grew during the seventies. But in the early eighties, the bubble burst. Softening minerals demand collided with rapid production expansion to create an unprecedented, sustained drop in mineral prices. From 1980 to 1985, the IMF index of non-fuel mineral prices fell nearly 30 percent. The decline was exacerbated by the tendency of nationalized mines to continue producing in large quantities despite falling prices. While private producers cut back their output, some nations, such as Zaire, Peru, and Zambia, had become too dependent on mineral revenue to reduce production; a few, such as Chile, even expanded production in an attempt to make up for lost revenues, and prices were driven even lower. Chilean copper output grew by nearly 30 percent between 1981 and 1985, while the price of the metal plunged by more than 20 percent.[59]

The overall result of these developments was a dramatic transformation of the world mineral industry—from a relatively stable, lucrative oligopoly to an unpredictable, intensely competitive business. This change undermined overnight the development strategy followed by many Third World minerals producers. No longer, as the authors of a 1990 study put it, can mineral resources "be regarded as buried treasure." In other words, if mineral projects are hard pressed to pay for themselves, they cannot be expected to provide much help in economic development beyond the mine.[60]

The traditional response to this dilemma, followed by the United States and other industrialized nations with long mining histories, is to move into mineral processing and fabrication, which add more value to mineral products than earlier stages of production. Unfortunately, developing countries have found it difficult to do so. Trade barriers (tariffs are generally higher on refined or fabricated metals than on ores or concen-

"From 1980 to 1985, the IMF index of non-fuel mineral prices fell nearly 30 percent."

trates), the reluctance of many private firms to assist in activities that may affect the markets for their own products, and other factors have kept Third World mineral producers from moving heavily into the higher-value steps of the production process.[61]

It may be time, instead, for developing countries to turn—as a growing grassroots development movement now suggests—to an alternative economic strategy, focused less on generation of export revenue and more on activities that directly reduce poverty and its human costs. These include promotion of grassroots, rural development and local agricultural production, education, health care, and other basic human needs. The outlines of such an approach are still vague, but it is clear that, for many Third World countries, and particularly those in Africa and Latin America, the current strategy of export-led development has failed.

Cleaning Up

Since the time of Agricola, the destruction from mineral production has been justified in the name of human progress. The sixteenth century scholar was quite conscious of mining's effects on the environment, yet argued that, without metals, "men would pass a horrible and wretched existence in the midst of wild beasts." In an absolute sense, Agricola was right: civilizations have always depended heavily on minerals for survival—and still do.[62]

But as Lewis Mumford put it in his classic *Technics and Civilization*, "One must admit the devastation of mining, even if one is prepared to justify the end....What was only an incidental and local damage in [Agricola's] time became a widespread characteristic of Western Civilization just as soon as it started in the eighteenth century to rest directly upon the mine and its products." Mining's effects on the earth, far from being merely local as often depicted, are now on the same scale as such hugely destructive natural forces as erosion.[63]

Cleaning up the mineral industry and its legacies will not be easy. Perhaps the hardest task will be to clean up abandoned mineral projects, because doing so often requires moving, treating, and containing extraordinarily large amounts of material spread over large areas. For example, about 7 million tons of tailings are present at the Eagle Mine Superfund site in Gilman, Colorado, and more than 200 million cubic meters of materials are stored in the 3,500 hectares of tailings ponds in the Clark Fork area of Montana. The latter contain an estimated 200 tons apiece of cadmium and silver, 9,000 tons of arsenic, 20,000 tons of lead, 90,000 tons of copper, and 50,000 tons of zinc.[64]

The consequences of not cleaning up such abandoned operations can be severe and long-lasting. The very reason these sites will be difficult to clean up is the most compelling reason to do so, since such huge volumes of material can cause abandoned mines to continue paying negative dividends—in such forms as sediment-choked streams, acid drainage, and metal contamination—for centuries.[65]

Aside from technical challenges, however, the chief problem in cleaning up old mines is that the mine operators are gone—and with them, the money for cleanup. Hence, governments often end up with the bill, which can be huge. The price tag for the Clark Fork cleanup, for example, is estimated to be $1 billion. Furthermore, the number of sites to be evaluated for cleanup is quite large: EPA estimates that between 800 and 1,500 mining sites need assessment, and that 70-100 will require remedial action. Nations with similar mining histories are likely to have similar cleanup needs.

The United States has chosen to fund mine cleanups through its Superfund program, which covers all types of abandoned industrial facilities. The program is primarily funded by a tax on chemical feedstocks. Progress on mineral sites has been sluggish, however. The special scale and characteristics of such sites may merit establishment of a separate program for former mineral facilities. One possible approach would be to fund mine cleanups through taxes on virgin mineral consumption. A similar program already exists for abandoned coal mines.[66]

Such taxes would serve another purpose as well, since they would also help create incentives for more efficient use of minerals, reducing the need for new mines. Some of these funds could be directed to Third World countries for cleanup of their abandoned mines. Without outside assistance, it is unlikely that many developing countries will be able to afford cleanups of their old mineral sites, which in most cases were developed for export to richer nations.

New technologies may help reduce the costs of cleanup. The high metal content of some tailings—which can pose the threat of contamination—can be turned to advantage in mine cleanups if methods are available for extracting the remaining metals. Thus, reprocessing old tailings can sometimes not only help reduce environmental hazards at a site, but also yield a salable product to help pay for cleanup.

Such methods are not only useful at abandoned sites, but also may help in reducing pollution at operating mines. Biological leaching—in which bacteria are used to extract metal from ore—is a promising new method. At the Los Bronces copper mine in Chile, a biological extraction project now being put into place is designed to avoid pollution in the Mantaro River, the source of Santiago's drinking water, by extracting copper from water repeatedly circulated through waste, overburden, and marginal ore dumps. The project's designers believe it will eventually recover more than a half-million tons of pure copper. An even more ambitious project at La Escondida, another giant Chilean copper mine, will recover pure metal from ore without smelting. Instead, copper will be extracted from concentrates by an ammonia solution, and then precipitated by electrolysis. In addition to avoiding the pollution and expense of a smelter, the project will have the advantage of producing pure copper rather than less-valuable concentrates.[67]

A variety of other practices can help reduce environmental impacts at every stage of the mineral extraction process. While destruction of surface features—be they forests or villages—is usually inevitable with surface mining, a variety of techniques can help cut down air, water, and soil pollution, and sometimes can return mined land to stable (if not its original) condition. In the initial excavation and mining process, careful storage of topsoil can ensure its availability for reclamation after mining is finished. If soil and rock are stored in well-designed impoundments, runoff and sedimentation problems can be kept to a minimum. Similarly, more careful storage and disposal of tailings can minimize the opportunity for them to contaminate their surroundings. Air pollution controls can substantially reduce emissions from smelters. Advanced methods—especially biological techniques—for extracting metals from ore could also offer substantial energy savings if they replace thermal (smelting) or mechanical methods.

Such careful attention by mine operators to minimizing environmental damage is unlikely, however, unless governments have the resources—and political will—to require it. While the mineral sector is currently subject to a broad range of environmental rules in most industrial and some developing countries, legal loopholes, lack of government funds, and weak enforcement are still allowing the creation of new environmental disaster areas.

For example, in the United States, which is widely regarded as a leader in such regulation, smelter emissions are regulated under the Clean Air Act and mining-caused water pollution under the Clean Water Act. Unfortunately, federal regulation of mining itself remains quite weak. EPA has done little to regulate disposal of mining wastes, despite their status as the single largest category of waste produced. In the 1980 Bevill Amendment to the Resource Conservation and Recovery Act of 1976, Congress exempted most mining wastes from regulation as hazardous waste, pending an EPA determination of their status. EPA has since decided to retain the exemption for most types of mineral industry waste, though final rules are still in process. In general, however, the states play a more important role in mining regulation, and the level of attention and enforcement varies dramatically.[68]

Nonetheless, industrial nations have available to them the funding and government personnel to put in place and more effectively enforce environmental laws for mineral producers—if the political will exists to do so. In the United States—by far the largest mineral producer among the industrial market nations—lawmakers now have a chance to strengthen environmental provisions in several laws affecting the impacts of mining, including the Resource Conservation and Recovery Act, the Clean Water Act, and the General Mining Act. Amendments to the Emergency Planning and Community Right-to-Know Act now pending would require the mining industry to report its toxic emissions to state and federal regulators.[69]

For developing countries, the challenge is much greater. While many of them have broad environmental protection laws, specific regulation of the minerals industries is rare. Where environmental laws do exist, funding and staff for enforcement are usually scarce. Chile, for example, has comprehensive and stringent environmental rules for mining, but they are virtually unenforced. The Chilean government has been particularly loath to force state-owned mineral operations to comply with the laws. Other countries with large state mining companies face similar conflicts of national interest between state mining companies and regulators, with local people and the environment most often the losers.[70]

At times, the prospect of major revenue from projects leads government officials to simply ignore environmental rules or studies. At the Ok Tedi copper and gold mine in Papua New Guinea, the government allowed the project's operators—an international consortium of private firms—to dump up to 150,000 tons of tailings a day into the nearby Fly River rather than contain them at the mine site, despite studies showing the potential for major damage to the river system.[71]

The pressure to neglect environmental concerns in favor of continued mineral output will continue to be strong unless broad changes occur in development policies and international debt service requirements. A legislative foundation already exists in some developing countries, however, for improved regulation of mineral industries. Several major mineral producers, including Chile, Brazil, and Peru, have recently started looking at mineral production's impacts on the environment and are attempting to improve the regulation of such activities.[72]

Additional pressure for environmental improvements could be created through substantial international assistance for environmental regulation and enforcement, as well as through the attachment of environmental conditions to mineral development funding. A portion of virgin mineral taxes levied in industrial countries could be allocated to improving the capacity of mineral-exporting developing countries to regulate their industries. Loans from development banks and their affiliates could include substantial components earmarked specifically for environmental protection, as non-governmental organizations in both industrialized and developing countries have urged in recent years. Some progress is already being made in this direction. The International Finance Corporation (the World Bank affiliate that lends to private-sector projects) has begun insisting on environmental impact assessments for mineral projects it funds. Not until environmental concerns play a major role in the decisions on whether and how to fund projects, however, will the situation improve dramatically.

In the short run, better regulation of the environmental impacts of mining is the most obvious way to reduce the damage done in supplying the world with minerals. There is considerable room for improvement of current mining practices—in increased attention to environmental safeguards, more sensitivity to local people and their concerns, and better planning for the indirect impacts of mineral development. More attention and new approaches to reducing the environmental impacts of currently operating mineral facilities could help to prevent the creation of more abandoned sites in the future. Operations can be made even more benign if they are designed with environmental concerns in mind from the outset. Additional research, such as that of the Mining and Environment Research Network—a far-flung group of independent reseachers investigating the impacts of mineral extraction in developing countries—should ease the task of cleaning up the mineral industries.[73]

In the long run, however, the benefits to be gained through mining regulation, while critically important, are not enough. Even well-managed mines are often enormously destructive. Careful reclamation may reduce erosion and pollution problems at mine sites, but ecological complexity and high costs usually preclude restoring the land to its previous condition. High energy use in mining and smelting makes reuse and recycling of metal-containing products almost always preferable to virgin production. To dramatically reduce the impacts of the minerals industry, attention must be paid not only to the extraction process, but to how mineral products are used.

Digging Out

The ultimate solution to the problem of mining's environmental destruction will require profound changes in both minerals use and in the global economy. No country has yet developed and put into place comprehen-

"It is the extraction and processing of minerals, not their use, that poses the greatest threat."

sive policies on the use of minerals and other raw materials. The assumption that prosperity is synonymous with the quantities of minerals taken from the earth has shaped the industrial development strategies of both capitalist and socialist nations. But that assumption is open to question. The environmental damage from nonstop growth in mineral production will eventually outweigh the benefits of increased materials supplies—if it does not already.

The way out of the trap lies in a simple distinction: it is the extraction and processing of minerals, not their use, that poses the greatest threat. The de facto materials policies of industrial nations have always been to champion the production of virgin minerals. Although such an approach has effectively promoted mining, it has also helped make minerals artificially cheap. This has led to widespread waste of mineral products, and has diverted funds that might have been used more productively to serve other needs.

A far less destructive policy would be to maximize conservation of mineral stocks already circulating in the global economy, thereby reducing both the demand for new materials and the environmental damage done to produce them. The world's industrial nations, the leading users of minerals, offer the most obvious opportunities for cutting demand. Minerals use in those nations is still rising, but increases have been slower in the last two decades than before. A growing body of evidence suggests that per capita needs for virgin minerals have already peaked there, and that major shifts are underway in the mix of minerals needed.[74]

National governments could accelerate the transition to more materials-efficient economies through basic changes in policies that govern the exploitation and use of raw materials. Tax policy offers the most obvious tool with which to start. Taxing, rather than subsidizing, production of virgin minerals would create stronger incentives to use them more efficiently. It could also provide governments with a way of paying for mine cleanups, as well as augmenting general revenues.

Many technical possibilities exist for using minerals more efficiently. The most obvious is recycling, as there is ample room to increase recycling rates for many metals. A 1992 U.S. Bureau of Mines study found that 10.6 million tons of iron and steel, 800,000 tons of zinc, and 250,000 tons of copper are discarded in U.S. solid waste each year. Though they recycle 45 percent of the aluminum they use, U.S. residents still throw away so much of the metal each year—2.3 million tons—that the energy saved by recycling it could meet the annual electricity needs of a city the size of Chicago.[75]

Beyond recycling, however, even more opportunity lies in making mineral-containing products more durable and repairable. More than a decade ago, a study by the U.S. Office of Technology Assessment concluded that reuse, repair, and remanufacturing of metal-containing products were the most promising methods of conserving metals. Governments could promote such practices by requiring manufacturers to offer longer warranties or to take products back at the end of their useful lives. Deposit/refund systems, for items as diverse as beverage containers and automobiles, can encourage consumers to return products for reuse instead of throwing them away.[76]

A particularly promising initiative has been undertaken by several European auto manufacturers, including BMW, Mercedes-Benz, Peugeot, Renault, Volkswagen/Audi, and Volvo, to make their vehicles entirely and easily recyclable. Engineers at the firms are designing cars with an eye toward easy disassembly, reuse, and recycling of various parts, and are attempting to minimize the use of nonrecyclable or hazardous materials. The approach could easily be adopted for other products.[77]

Another option is to substitute more benign materials for those whose production is judged to be the most environmentally damaging. Such judgments are inherently difficult, since comparison of the environmental impacts of different materials is an inexact science. But some minerals stand out from the crowd. Production of copper, for example, is exceptionally destructive. The use of optical fibers made of glass, in place of copper wires used in communications, is an encouraging example of a shift to a less-damaging substitute. Fiber optics also offer a much greater information-carrying capacity than copper wire. Similarly, the large amounts of energy required for aluminum production make it a logical candidate for replacement with other materials in applications where its light weight does not compensate by saving even greater amounts of energy than its production requires. The energy taxes now proposed as measures to reduce carbon emissions would speed shifts to less energy-intensive materials.

More difficult than shifting industrial nations to a minerals-efficient economy will be the search for a path to a sustainable future for developing countries. For those now heavily dependent on mineral exports, a rapid decline in demand could have dire consequences. Development planners need to recognize that mineral projects have generally failed to deliver long-term national economic success, and that current trends in minerals markets and use make mining an unpromising sector for future investment. More attention—and funding—needs to be devoted to diversifying the economies of mineral-producing nations and regions. Industrial nations will need to consider devoting a substantial share of any taxes levied on virgin minerals to development assistance to producing nations. One prerequisite for successfully rehabilitating those economies is relief from the crushing burden of international debt.

The greatest challenge, however, lies in the search for a Third World development strategy that allows poor countries to improve human welfare dramatically without using and discarding hundreds of kilograms of metals and other minerals per person each year. Roughly three fourths of the world's people now live in countries that generally have yet to build the railways, roads, bridges, buildings, and other basic infrastructure so essential to rich nations' economies. These countries will inevitably need more minerals as their development proceeds.

But basic choices in development plans—in the location, scale, density, and character of urban development, for example, or in the types of transportation systems emphasized—could have dramatic impacts on the amount and mix of materials needed, and on the environmental costs of producing them. It is critical that development planners take minerals use into account in the same way that they consider energy use, water use, and the production of food. The urgency of this issue cannot be overestimated: if more countries follow the development path of the rich industrial nations, world demand for minerals will continue to rise for many years. Given current geographical trends in production, it is the Third World itself that will suffer most of the environmental damage from meeting the demand.

In the past decade, concerns about the declining quality of the environment have brought anguished reexaminations of virtually all the major economic activities on which civilization depends. Nearly all—from energy production to transportation, from forestry to agriculture—have been called upon to reduce their toll on the natural systems that underpin the global economy. Despite this, mineral production—perhaps because of its remoteness or because of its growing concentration in countries that depend on it so heavily—has been relatively free of scrutiny. Yet this sector, on which so many others depend, is one of the largest users of energy, despoilers of air, water, and land, and producers of industrial waste, and therefore one of the most desperately in need of reform.

With analysis of the environmental impacts of other industries has come a growing realization that prevention is better than cure—that the greatest environmental benefits are usually yielded by basic alterations in production processes and consumption patterns, rather than through "pollution control." Mining's devastating environmental effects make the ultimate case for a strategy emphasizing pollution prevention, and not just control. As an inherently destructive industry that supplies raw materials throughout the economy, its impacts are best reduced by basic changes in *other* industries, and in the societies that eventually use mineral products.

While mining companies and the governments of nations heavily dependent on mining exports may feel the costs of such a strategy are unacceptable, we have now passed the point where we can continue to live with anything less than a full accounting for today's policies. Reducing the global appetite for minerals, and the environmental toll of the huge industries that satisfy it, will not be easy. But allowing the devastation to grow unchecked could prove to be an even more difficult—and costly—course for humanity.

Notes

1. Mining projects in national parks from Olga Sheean "Fool's Gold in Ecuador," *World Wildlife Fund News*, January/February 1992.

2. U.S. Library of Congress, Congressional Research Service, *Are We Running Out? A Perspective on Resource Scarcity* (Washington, D.C.: U.S. Government Printing Office (GPO), 1978).

3. Rex Bosson and Bension Varon, *The Mining Industry and the Developing Countries* (New York: Oxford University Press, 1977).

4. Increase in iron, copper, and zinc production derived from historical data in Bosson and Varon, *The Mining Industry*, and from 1991 production figures in U.S. Bureau of Mines (USBM), *Mineral Commodity Summaries 1992* (Washington, D.C.: 1992); history of aluminum from John A. Wolfe, *Mineral Resources: A World Review* (New York: Chapman and Hall, 1984); aluminum production from USBM, *Mineral Commodity Summaries.*

5. Uranium is a metal, but is classed here with the fossil fuels to distinguish it from the nonfuel minerals analyzed in this paper.

6. Information on uses from USBM, *Mineral Commodity Summaries.*

7. Aluminum and steel prices and use of metals in steelmaking from USBM, *Mineral Commodity Summaries;* estimated relative value of world metals sales is a Worldwatch estimate, based on data in ibid.

8. 1990 consumption figures derived from data in World Resources Institute (WRI), *World Resources 1992-93* (New York: Oxford University Press, 1992); industrial nations' historical shares of steel and other metals consumption from Olivier Bomsel et al., *Mining and Metallurgy Investment in the Third World: The End of Large Projects?* (Paris: Organisation for Economic Co-operation and Development (OECD), 1990).

9. Minerals demand from Bomsel et al., *Mining and Metallurgy Investment.*

10. For more on the materials use trends discussed in this section, see Bomsel et al., *Mining and Metallurgy Investment;* Marc H. Ross and Robert H. Williams, *Our Energy: Regaining Control* (New York: McGraw-Hill, 1981); Eric D. Larson et al., "Materials, Affluence, and Industrial Energy Use," *Annual Review of Energy, Vol. 12* (Palo Alto, Calif.: 1987); Peter F. Drucker, "The Changed World Economy," *Foreign Affairs*, Spring 1986; Robert U. Ayres, "Industrial Metabolism," and Robert Herman et al., "Dematerialization," in Jesse H. Ausubel and Hedy E. Sladovich, eds., *Technology and Environment* (Washington, D.C.: National Academy Press, 1989); and USBM, *The New Materials Society, Volume 3: Materials Shifts in the New Society* (Washington, D.C.: 1991).

11. World recycling of aluminum from United Nations Environment Programme, *Environmental Data Report 1991/92* (Cambridge, Mass.: Basil Blackwell, 1991).

12. USBM, *The New Materials Society.*

13. Bomsel et al., *Mining and Metallurgy Investment.*

14. USBM, *Mineral Commodity Summaries.*

15. Import dependence of Japan and Western Europe from Bosson and Varon, *The Mining Industry*, and from Faysal Yachir, *Mining in Africa Today* (London: Zed Books, 1988); United States from USBM, *Mineral Commodity Summaries.*

16. Yachir, *Mining in Africa Today.*

17. Reserve figures from WRI, *World Resources 1992-93;* growth of reserves from "Human Factors Influencing Resource Availability and Use: Group Report," in Digby J. McLaren and Brian J. Skinner, eds., *Resources and World Development* (London: John Wiley and Sons, 1987).

18. Reserves from USBM, *Mineral Commodity Summaries;* István Dobozi, "Perestroika and the End of the Cold War: Possible Mineral Trade Implications for the USSR and Eastern Europe," presentation at annual meeting of the American Association for the Advancement of Science, Washington, D.C., February 17, 1991.

19. Georgius Agricola, *De Re Metallica* (New York: Dover Publications, 1950).

20. Trucks and shovels from Bosson and Varon, *The Mining Industry.*

21. The sources for Table 4 are: Julio Díaz Palacios, "Environmental Destruction in Southern Peru," *Earth Island Journal*, Summer 1989; Nauru from "Who Will Clean Up Paradise," *Asiaweek*, January 4, 1991, and Martin Weston, Nauru Government Office, letter to the editor, *Economist*, February 23, 1991; Philip M. Fearnside, "The Charcoal of Carajás: A Threat to the Forests of Brazil's Eastern Amazon Region," *Ambio*, Vol. 18, No. 2, 1989; Severonikel from Valeriy E. Berlin, "Conservation Efforts in Kola Peninsula's Lapland Preserve Detailed," *JPRS Reports*, June 17, 1991; "Sabah Mining Pollution - Part One: Villagers Demand US $6 Million Compensation," APPEN (Asia-Pacific Peoples Environment Network) Features, Penang, Malaysia, 1990; David Cleary, *Anatomy of the Amazon Gold Rush* (Iowa City: University of Iowa Press, 1990). Superfund sites from Steve Hoffman, U.S. Environmental Protection Agency (EPA), Washington, D.C., private communication, November 5, 1991; EPA and Montana Department of Health and Environmental Sciences (MDHES), *Clark Fork Superfund Master Plan* (Helena, Mont.: 1988); Peter Nielsen and Bruce Farling, "Hazardous Wastes Endanger Water, Wildlife, Land: Mining Catastrophe in Clark Fork," *Clementine* (Mineral Policy Center, Washington, D.C.), Autumn 1991.

22. EPA and MDHES, *Clark Fork Superfund;* Nielsen and Farling, "Mining Catastrophe in Clark Fork"; Peter Nielsen, Executive Director, Clark Fork Coalition, Missoula, Mont., private communication, October 16, 1991; Cabinet Mountains from "Shame on Montana" (videotape), World Wide Film Expedition, Missoula, Mont., 1991.

23. USBM, *1989 Minerals Yearbook* (Washington, D.C.: GPO, 1990).

24. Share of metal ore discarded as tailings, metal contaminants in tailings, and tailings pond examples from Johnnie N. Moore and Samuel N. Luoma, "Large-Scale Environmental Impacts: Mining's Hazardous Waste," *Clementine* (Mineral Policy Center, Washington, D.C.), Spring 1991; sulfur content of metal ores from Martyn Kelly, *Mining and the Freshwater Environment* (London: Elsevier Science Publishers, 1988); organic contaminants from Christopher G. Down and John Stocks, *Environmental Impact of Mining* (New York: John Wiley and Sons, 1977); toluene use in concentrators from Daniel M. Horowitz, "Mining and Right-to-Know," *Clementine*, Winter 1990; health effects of toluene from Eric P. Jorgenson, ed., *The Poisoned Well: New Strategies for Groundwater Protection* (Washington, D.C.: Island Press, 1989).

25. Michael C. Howard, *Mining, Politics, and Development in the South Pacific* (Boulder, Colo.: Westview Press, 1991); David Hyndman, "Digging the Mines in Melanesia," *Cultural Survival Quarterly*, Vol. 15, No. 2, 1991; David Clark Scott, "Rebels Keep Papua New Guinea Mine Closed, " *Christian Science Monitor*, July 7, 1989; "A Mine of Controversy," *South*, June/July 1991; Moore and Luoma, "Mining's Hazardous Waste"; Don Hinrichsen, *Our Common Seas* (London: Earthscan, 1990).

26. Detlev Möller, "Estimation of the Global Man-Made Sulphur Emission," *Atmospheric Environment*, Vol. 18, No. 1, 1984; effects of fluoride emissions from Paul R. Ehrlich et al., *Ecoscience: Population, Resources, Environment* (San Francisco, Calif.: W.H. Freeman, 1977).

27. Dead zones from Moore and Luoma, "Mining's Hazardous Waste"; Sudbury fish kills and Trail smelter pollution from Down and Stocks, *Environmental Impact of Mining*.

28. Berlin, "Conservation Efforts in Kola Peninsula's Lapland Preserve Detailed"; Chilean and U.S. smelter emissions from United Nations, Economic Commission for Latin American and the Caribbean, *Sustainable Development: Changing Production Patterns, Social Equity and the Environment* (Santiago, Chile: 1991).

29. Historical grade of copper ore from Bosson and Varon, *The Mining Industry*.

30. Scale of Bingham Canyon mine from Andrew Goudie, *The Human Impact on the Natural Environment* (Cambridge, Mass.: MIT Press, 1990); Kennecott toxics report from Horowitz, "Mining and Right-to-Know."

31. Goldstrike from Kenneth Gooding, "American Barrick's Glittering Run of Luck Continues," *Financial Times*, October 4, 1991; Amazon mining from Cleary, *Anatomy of the Amazon Gold Rush*; hydraulic mining in California from Duane A. Smith, *Mining America: The Industry and the Environment, 1800-1980* (Lawrence: University Press of Kansas, 1987).

32. Gold price and mercury in Amazon from Cleary, *Anatomy of the Amazon Gold Rush*; gold rush in Indonesia from Alexander Gurov, "Gold Rush in Kalimantan," *Asia and Africa Today*, No. 2, 1990; Zimbabwe from Paul Jourdan, Institute of Mining Research, Harare, Zimbabwe, private communication, April 12, 1991; other nations from Melvyn Westlake and Robin Stainer, "Rising Gold Fever," *South*, March 1989; mercury in Madeira River from Jerome A. Nriagu et al., "Mercury Pollution in Brazil," *Nature*, April 2, 1992.

33. "Heavy Rains Burst South Carolina Dam: Major Cyanide Spill," *Clementine* (Mineral Policy Center, Washington, D.C.), Winter 1990; bird kills from Alyson Warhurst, "Environmental Degradation from Mining and Mineral Processing in Developing Countries: Corporate Responses and National Policies," draft discussion document for meeting of the Mining and Environment Research Network, Steyning, United Kingdom, April 10-13, 1991.

34. Total world mineral extraction and waste generated are Worldwatch Institute estimates, based on production data in USBM, *Mineral Commodity Summaries*, and average ore grades in Donald G. Rogich, "Trends in Material Use: Implications for Sustainable Development," unpublished paper, U.S. Bureau of Mines, Division of Mineral Commodities, 1992; the estimated sediment load in the world's rivers is 16.5 billion tons/year, according to J.D. Milliman and R.H. Meade, cited in Brian J. Skinner, "Resources in the 21st Century: Can Supplies Meet Needs?" *Episodes*, December 1989.

35. Area mined is a Worldwatch estimate, derived by multiplying world production data by 1980 U.S. land use/production ratios; world production from USBM, *Mineral Commodity Summaries*; ratios derived from Wilton Johnson and James Paone, "Land Utilization and Reclamation in the Mining Industry, 1930- 80," Bureau of Mines Information Circular 8862, Washington, D.C., 1982; Elizabeth Dore, "Open Wounds," *NACLA Report on the Americas*, September 1991.

36. Sussex from Down and Stocks, *Environmental Impact of Mining*; Comstock Lode and observations in 1877 by William Wright from Smith, *Mining America*.

37. Fearnside, "The Charcoal of Carajás."

38. Grande Carajás from Dore, "Open Wounds"; Albrás from Bomsel et al., *Mining and Metallurgy Investment*; Albrás production capacity from "A Month to Assess Albrás Damage," *Gazeta Mercantil*, March 18, 1991; Tucuruí electricity used by Albrás from Liliana Acero, Centro de Investigacion y Promocion Educativa y Social, Buenos Aires, private communication, May 7, 1992; for information on the destructive nature of the Tucuruí dam, see Barbara J. Cummings, *Dam the Rivers, Damn the People* (London: Earthscan, 1990); electricity use of Brazilian aluminum industry from Howard Geller, *Efficient Electricity Use: A Development Strategy for Brazil* (Washington, D.C.: American Council for an Energy-Efficient Economy, 1991).

39. Energy requirements of aluminum smelting from U.S. Congress, Office of Technology Assessment (OTA), *Nonferrous Metals: Industry Structure—Background Paper* (Washington, D.C.: GPO, 1990); requirements for mining, beneficiation, and alumina refining from Martin Brown and Bruce McKern, *Aluminium, Copper, and Steel in Developing Countries* (Paris: OECD, 1987); total energy consumption in aluminum production is a Worldwatch estimate derived from these sources and from 1991 production estimate in USBM, *Mineral Commodity Summaries*; electricity use in smelting converted into GJ assuming a 64/36 percent mix of fossil and non-fossil electricity generation (while a large portion of aluminum smelting is powered by hydroelectric sources, much of that energy could be directed into other uses now dependent on fossil energy sources); electricity use in Africa from U.S.

Department of Energy (DOE), Energy Information Administration, *International Energy Annual 1990* (Washington, D.C.: 1992); low electricity rates from Merton J. Peck, ed., *The World Aluminum Industry in a Changing Energy Era* (Washington, D.C.: Resources for the Future, 1988) and Ronald Graham, *The Aluminium Industry and the Third World* (London: Zed Books, 1982); for additional information on the environmental impacts of the aluminum industry, see John E. Young, "Aluminum's Real Tab," *World Watch*, March/April 1992.

40. Energy use in copper production is a Worldwatch estimate derived from 1990 production estimate in USBM, *Mineral Commodity Summaries*, and from estimates of energy requirements per ton of copper in Brown and McKern, *Aluminum, Copper, and Steel*, and in Bernard A. Gelb and Jeffrey Pliskin, *Energy Use in Mining: Patterns and Prospects* (Cambridge, Mass.: Ballinger Publishing, 1979); energy use in steelmaking is from DOE, Energy Information Administration, *Manufacturing Energy Consumption Survey: Consumption of Energy 1988* (Washington, D.C.: 1991); figure includes some energy used in fabrication as well as crude production, but does not include that used in mining and ore concentration; minerals industry share of world energy use is a Worldwatch estimate based on information in sources above.

41. OTA, *Nonferrous Metals*; Brown and McKern, *Aluminium, Copper, and Steel.*

42. Howard, *Mining, Politics, and Development*; Hyndman, "Digging the Mines in Melanesia"; Scott, "Rebels Keep Papua New Guinea Mine Closed"; "A Mine of Controversy."

43. American Anthropological Association (AAA), "Report of the Special Commission to Investigate the Situation of the Brazilian Yanomami," Washington, D.C., 1991.

44. Quote from Richard Barnet, *The Lean Years: Politics in the Age of Scarcity* (New York: Simon and Schuster, 1980); AAA, "Report of the Special Commission"; decree from Comissão Pela Criação do Parque Yanomami, "Diario Oficial Publica Delimitação de Terra Indigena Yanomami em Area Continua," news bulletin, São Paulo, July 26, 1991, and Julia Preston, "Brazil Grants Land Rights to Indians," *Washington Post*, November 16, 1991.

45. U.S. miners' deductions must not exceed half the operation's taxable income; taxation of mineral industries from John J. Schanz, Jr., *The Subsidization of Non-Fuel Mineral Production at Home and Abroad* (Washington, D.C.: Congressional Research Service, 1987), from Talbot Page, *Conservation and Economic Efficiency, An Approach to Materials Policy* (Baltimore, Md.: Johns Hopkins University Press, 1977), and from National Commission on Supplies and Shortages, *Government and the Nation's Resources* (Washington, D.C.: GPO, 1976); special tax provisions for minerals are spelled out in United States Code, Vol. 26, sections 611-617; depletion allowances for various minerals are also listed in USBM, *Mineral Commodity Summaries.*

46. Tax subsidy to the mining industry from Executive Office of the President, *Budget of the United States Government* and *Special Analyses: Budget of the United States Government* (Washington, D.C.: GPO, various years); U.S. Congress, Joint Committee on Taxation,

"Estimates of Federal Tax Expenditures for Fiscal Years 1992-1996," Washington, D.C., GPO, April 1991.

47. For General Mining Act, see "Mining Reform Alternatives Compared: Point- by-Point," *Clementine* (Mineral Policy Center, Washington, D.C.), Spring/Summer 1990, and U.S. General Accounting Office (GAO), *Federal Land Management: The Mining Law of 1872 Needs Revision* (Washington, D.C.: 1989); lack of revenue and value of mineral production on federal land from James Duffus III, Director, Natural Resources Management Issues, GAO, testimony before the Subcommittee on Mining and Natural Resources, Committee on Interior and Insular Affairs, U.S. House of Representatives, Washington, D.C., September 6, 1990.

48. W.C.J. van Rensburg, *Strategic Minerals* (Englewood Cliffs, N.J.: Prentice-Hall, 1986).

49. Ibid.; GAO, "Federal Encouragement of Mining Investment in Developing Countries for Strategic and Critical Materials Has Been Only Marginally Effective," Washington, D.C., 1982.

50. British/German arms race from Paul Kennedy, *The Rise and Fall of the Great Powers* (New York: Random House, 1987); for the influence of military demand on the evolution of the aluminum industry, see Graham, *The Aluminum Industry*; quote from Harry N. Holmes, *Strategic Minerals and National Strength* (New York: MacMillan, 1942).

51. Page, *Conservation and Economic Efficiency*.

52. David J. Fox, "Bolivian Mining, a Crisis in the Making," in Thomas Greaves and William Culver, eds., *Miners and Mining in the Americas* (Manchester, U.K.: Manchester University Press, 1985).

53. World Bank, *World Debt Tables 1990-91: Supplement* (Washington, D.C.: 1991); average ratio excludes Suriname, because no debt data were available; average ratio calculated using most recent data available; all figures used were 1990 estimates, with the exception of Botswana, Zambia, and Guyana (1989), and Liberia (1987).

54. Economic decline in Zambia from Jane Perlez, "Rainy Days in Zambia (Price an Umbrella!)" *New York Times*, June 5, 1990, and from Tony Hodges, "Zambia's Autonomous Adjustment," *Africa Recovery* (United Nations, New York), December 1988; child deaths from Alan B. Durning, *Poverty and the Environment: Reversing the Downward Spiral*, Worldwatch Paper 92 (Washington, D.C.: Worldwatch Institute, November 1989); Zambian debt from World Bank, *World Debt Tables*.

55. Bomsel et al., *Mining and Metallurgy Investment*.

56. Nationalization in Guinea from Graham, *The Aluminum Industry*; all others from Bosson and Varon, *The Mining Industry*; cartels from ibid.

57. The mineral-centered development strategy described here is thoroughly laid out in

Bosson and Varon, *The Mining Industry*; World Bank price forecasts and Guelbs project from Bomsel et al., *Mining and Metallurgy Investment*.

58. Les Guelbs from Bomsel et al., *Mining and Metallurgy Investment*; GDP of Mauritania from World Bank, *World Development Report 1981* (New York: Oxford University Press, 1981).

59. Bomsel et al., *Mining and Metallurgy Investment*; USBM, *Minerals Yearbook* (various years); IMF price index from Ximena Cheatham, International Monetary Fund, private communication, October 1991; Chilean copper output and price decline from Carlos Fortin, "Chilean Copper Policy: International and Internal Aspects," *IDS Bulletin* (Institute of Development Studies, University of Sussex, Brighton, United Kingdom), October 1986.

60. Bomsel et al., *Mining and Metallurgy Investment*.

61. The U.S. tariffs for various mineral products are listed in USBM, *Mineral Commodity Summaries*.

62. Agricola, *De Re Metallica*.

63. Lewis Mumford, *Technics and Civilization* (New York: Harcourt Brace Jovanovich, 1963).

64. Eagle Mine from EPA, Office of External Affairs, Region VIII, "Mining Wastes in the West: Risks and Remedies—Overview," Denver, Colorado, 1987; Clark Fork tailings from Moore and Luoma, "Mining's Hazardous Waste."

65. Moore and Luoma, "Mining's Hazardous Waste."

66. The Surface Mining Control and Reclamation Act of 1977 established an Abandoned Mine Reclamation Fund, financed by a tax on coal production; Warren Freedman, *Federal Statutes on Environmental Protection: Regulation in the Public Interest* (New York: Quorum Books, 1987).

67. Alyson Warhurst, "Environmental Management in Mining and Mineral Processing in Developing Countries," *Natural Resources Forum*, February, 1992; Leslie Crawford, "Chile's Giant Copper Mine to Boost Production," *Financial Times*, February 28, 1992.

68. OTA, *Managing Industrial Solid Wastes From Manufacturing, Mining, Oil and Gas Production, and Utility Combustion* (Washington, D.C.: 1992).

69. The Senate Environment and Public Works Committee voted to approve amendments to the right-to-know law on April 30, 1992; the proposal still awaits action on the Senate floor and in the House of Representatives; "Key Vote Advances Right-to-Know-More," *Working Notes on Community Right-to-Know* (Working Group on Community Right-to-Know, Washington, D.C.), May 1992.

70. Chile from Warhurst, "Environmental Degradation from Mining."

71. Ok Tedi from Howard, *Mining, Politics, and Development.*

72. Warhurst, "Environmental Degradation from Mining"; "Chile Mining, Environment Leaders to Talk," *Journal of Commerce,* May 6, 1992.

73. The Mining and Environment Research Network consists of researchers in more than 20 countries, and is coordinated by Professor Alyson Warhurst of the Science Policy Research Unit, University of Sussex, Brighton, United Kingdom.

74. Minerals use trends from Bomsel et al., *Mining and Metallurgy Investment;* Ross and Williams, *Our Energy;* Larson et al., "Materials, Affluence, and Industrial Energy Use"; Drucker, "The Changed World Economy"; Ayres, "Industrial Metabolism"; Herman et al., "Dematerialization"; USBM, *The New Materials Society.*

75. Metals in U.S. solid waste from Rogich, "Trends in Material Use"; a ton of recycled aluminum takes about 15,700 kilowatt-hours less electricity to produce than a ton of primary aluminum; thus about 3.6 billion kilowatt- hours—about 1.3 percent of annual U.S. electricity use—could have been saved by recycling 2.3 million tons of aluminum; the population of Chicago is about 1.3 percent of that of the United States.

76. OTA, *Technical Options for Conservation of Metals: Case Studies of Selected Metals and Products* (Washington, D.C.: GPO, 1979).

77. "Old Cars Get a New Lease on Life," *Financial Times,* September 3, 1991; Bill Siuru, "Car Recycling in Germany," *Resource Recycling,* February 1991; "Peugeot Developing Facility to Recycle Junk Automobiles," *Multinational Environmental Outlook,* March 5, 1991; "Volvo Announces Plans to Recycle Cars as Part of Environmental Impact Scheme," *International Environment Reporter,* September 11, 1991; Krystal Miller, "On the Road Again and Again: Auto Makers Try to Build Recyclable Car," *Wall Street Journal,* April 30, 1991; "Daimler Has 10% of Recycler," *New York Times,* March 29, 1991; Stuart Marshall, "Green Scrapyards," *Financial Times,* March 23, 1991.

JOHN E. YOUNG is a Research Associate at the Worldwatch Institute, where his research focuses on materials, energy, and environmental pollution. He is the author of Worldwatch Paper 101, *Discarding The Throwaway Society,* and coauthor of three of the Institute's *State of the World* reports. He is a graduate of Carleton College, where he studied political science and technology policy.

THE WORLDWATCH PAPER SERIES

No. of
Copies

______ 57. **Nuclear Power: The Market Test** by Christopher Flavin.
______ 58. **Air Pollution, Acid Rain, and the Future of Forests** by Sandra Postel.
______ 60. **Soil Erosion: Quiet Crisis in the World Economy** by Lester R. Brown and Edward C. Wolf.
______ 61. **Electricity's Future: The Shift to Efficiency and Small-Scale Power** by Christopher Flavin.
______ 62. **Water: Rethinking Management in an Age of Scarcity** by Sandra Postel.
______ 63. **Energy Productivity: Key to Environmental Protection and Economic Progress** by William U. Chandler.
______ 65. **Reversing Africa's Decline** by Lester R. Brown and Edward C. Wolf.
______ 66. **World Oil: Coping With the Dangers of Success** by Christopher Flavin.
______ 67. **Conserving Water: The Untapped Alternative** by Sandra Postel.
______ 68. **Banishing Tobacco** by William U. Chandler.
______ 69. **Decommissioning: Nuclear Power's Missing Link** by Cynthia Pollock.
______ 70. **Electricity For A Developing World: New Directions** by Christopher Flavin.
______ 71. **Altering the Earth's Chemistry: Assessing the Risks** by Sandra Postel.
______ 73. **Beyond the Green Revolution: New Approaches for Third World Agriculture** by Edward C. Wolf.
______ 74. **Our Demographically Divided World** by Lester R. Brown and Jodi L. Jacobson.
______ 75. **Reassessing Nuclear Power: The Fallout From Chernobyl** by Christopher Flavin.
______ 76. **Mining Urban Wastes: The Potential for Recycling** by Cynthia Pollock.
______ 77. **The Future of Urbanization: Facing the Ecological and Economic Constraints** by Lester R. Brown and Jodi L. Jacobson.
______ 78. **On the Brink of Extinction: Conserving The Diversity of Life** by Edward C. Wolf.
______ 79. **Defusing the Toxics Threat: Controlling Pesticides and Industrial Waste** by Sandra Postel.
______ 80. **Planning the Global Family** by Jodi L. Jacobson.
______ 81. **Renewable Energy: Today's Contribution, Tomorrow's Promise** by Cynthia Pollock Shea.
______ 82. **Building on Success: The Age of Energy Efficiency** by Christopher Flavin and Alan B. Durning.
______ 83. **Reforesting the Earth** by Sandra Postel and Lori Heise.
______ 84. **Rethinking the Role of the Automobile** by Michael Renner.
______ 85. **The Changing World Food Prospect: The Nineties and Beyond** by Lester R. Brown.
______ 86. **Environmental Refugees: A Yardstick of Habitability** by Jodi L. Jacobson.
______ 87. **Protecting Life on Earth: Steps to Save the Ozone Layer** by Cynthia Pollock Shea.
______ 88. **Action at the Grassroots: Fighting Poverty and Environmental Decline** by Alan B. Durning.
______ 89. **National Security: The Economic and Environmental Dimensions** by Michael Renner.
______ 90. **The Bicycle: Vehicle for a Small Planet** by Marcia D. Lowe.

_____ 91. **Slowing Global Warming: A Worldwide Strategy** by Christopher Flavin.
_____ 92. **Poverty and the Environment: Reversing the Downward Spiral** by Alan B. Durning.
_____ 93. **Water for Agriculture: Facing the Limits** by Sandra Postel.
_____ 94. **Clearing the Air: A Global Agenda** by Hilary F. French.
_____ 95. **Apartheid's Environmental Toll** by Alan B. Durning.
_____ 96. **Swords Into Plowshares: Converting to a Peace Economy** by Michael Renner.
_____ 97. **The Global Politics of Abortion** by Jodi L. Jacobson.
_____ 98. **Alternatives to the Automobile: Transport for Livable Cities** by Marcia D. Lowe.
_____ 99. **Green Revolutions: Environmental Reconstruction in Eastern Europe and the Soviet Union** by Hilary F. French.
_____100. **Beyond the Petroleum Age: Designing a Solar Economy** by Christopher Flavin and Nicholas Lenssen.
_____101. **Discarding the Throwaway Society** by John E. Young.
_____102. **Women's Reproductive Health: The Silent Emergency** by Jodi L. Jacobson.
_____103. **Taking Stock: Animal Farming and the Environment** by Alan B. Durning and Holly B. Brough.
_____104. **Jobs in a Sustainable Economy** by Michael Renner.
_____105. **Shaping Cities: The Environmental and Human Dimensions** by Marcia D. Lowe.
_____106. **Nuclear Waste: The Problem That Won't Go Away** by Nicholas Lenssen.
_____107. **After the Earth Summit: The Future of Environmental Governance** by Hilary F. French.
_____108. **Life Support: Conserving Biological Diversity** by John C. Ryan.
_____109. **Mining the Earth** by John E. Young.

_____ **Total Copies**

☐ **Single Copy: $5.00**
☐ **Bulk Copies (any combination of titles)**
☐ 2–5: $4.00 each ☐ 6–20: $3.00 each ☐ 21 or more: $2.00 each

☐ **Membership in the Worldwatch Library: $25.00 (overseas airmail $40.00)**
The paperback edition of our 250-page "annual physical of the planet," *State of the World 1991,* plus all Worldwatch Papers released during the calendar year.

☐ **Subscription to *World Watch* Magazine: $15.00 (overseas airmail $30.00)**
Stay abreast of global environmental trends and issues with our award-winning, eminently readable bimonthly magazine.

No postage required on prepaid orders. Minimum $3 postage and handling charge on unpaid orders.

Make check payable to Worldwatch Institute
1776 Massachusetts Avenue, N.W., Washington, D.C. 20036-1904 USA

Enclosed is my check for U.S. $________

____

name **daytime phone #**

____

address

____

city **state** **zip/country**